The Truth About Your Food

Be Smart About Soft Drinks

By Rachael Morlock

Cavendish Square

Published in 2023 by Cavendish Square Publishing, LLC
2544 Clinton Street, Buffalo, NY 14224

Website: cavendishsq.com

Library of Congress Cataloging-in-Publication Data

Names: Morlock, Rachael, author.
Title: Be smart about soft drinks / Rachael Morlock.
Description: Buffalo, New York : Cavendish Square Publishing, [2023] |
Series: The truth about your food | Includes index.
Identifiers: LCCN 2022027005 | ISBN 9781502665881 (library binding) | ISBN 9781502665874 (paperback) | ISBN 9781502665898 (ebook)
Subjects: LCSH: Soft drinks–Juvenile literature. | Nutrition–Juvenile literature. | Food–Safety measures–Juvenile literature.
Classification: LCC TP630 .M67 2023 | DDC 663/.62–dc23/eng/20220808
LC record available at https://lccn.loc.gov/2022027005

Editor: Rachael Morlock
Copyeditor: Shannon Harts
Designer: Deanna Paternostro

The photographs in this book are used by permission and through the courtesy of: Cover (main) Ievgenii Meyer/Shutterstock.com; cover (DNA graphic), back cover, pp. 3, 4, 7, 11, 15, 18, 23, 32-33, 35, 37, 42-43, 44-45, 46-47, 48 Omelchenko/Shutterstock.com; pp. 5, 13 Brent Hofacker/Shutterstock.com; p. 6 Rawpixel.com/Shutterstock.com; p. 8 Luis Echeverri Urrea/Shuttersrtock.com; p. 11 Alla Simacheva/Shutterstock.com; p. 14 DuxX/Shutterstock.com; p. 17 GaudiLab/Shutterstock.com; p. 19 Pormezz/Shutterstock.com; p. 20 LightField Studios/Shutterstock.com; p. 22 Unchalee Khun/Shutterstock.com; p. 25 adriaticfoto/Shutterstock.com; p. 27 Magic mine/Shutterstock.com; p. 28 vipman/Shutterstock.com; p. 30 DigArt/Shutterstock.com; p. 33 antoniodiaz/Shutterstock.com; p. 34 AnikonaAnn/Shutterstock.com; p. 36 Maria Shipakina/Shutterstock.com; p. 37 Rattanasak Khuentana/Shutterstock.com; p. 39 Tada Images/Shutterstock.com; p. 40 Monkey Business Images/Shutterstock.com.

CPSIA compliance information: Batch #CW23CSQ: For further information, contact Cavendish Square Publishing LLC at 1-877-980-4450.

Printed in the United States of America

Contents

Introduction

You don't have to look far to find soft drinks. They're available as refreshments in most stores, restaurants, vending machines, and movie theaters. Soft drink is a general term for a drink that is flavored, sweetened, and usually **carbonated**. Carbonated soft drinks are often called coke, cola, pop, soda, or soda pop, depending on where you live. Root beer, ginger ale, cola, orange, lemon-lime, and cream are just a few of the many flavors of fizzy sodas. They come in a variety of colors and brands as well.

A version of a soft drink existed as early as the 1600s. It was made with water, lemon, and honey. In the next century, people began experimenting with carbonated drinks. Joseph Priestley designed a device for pumping carbon dioxide into water in the late 1700s. Others developed Priestley's techniques in order to mix, bottle, and sell carbonated water. This soda water was sold in the United States by 1807. Restaurants and pharmacies offered soda water mixed with sweet, flavored syrups to customers. Pharmacies even added drugs to these drinks and suggested that they had health benefits.

It was once a special treat to buy a soda, but soft drinks have become everyday beverages in the past few decades. Many Americans buy soda in large quantities and drink more than one serving a day. In 1998, soda consumption reached an all-time high. The average American drank a total of 53 gallons (200 liters) of soda that year! In recent years, soda consumption has dropped, but it's still high. Most children today drink about 30 gallons (113.6 L) of soda a year.

Powdered versions of lemonade, fruit drinks, and iced tea that you mix with water are considered soft drinks, but they aren't carbonated.

Soft drinks are also known as sugar-sweetened beverages. These drinks have sugar added to them, and they include soda, fruit drinks, sports drinks, energy drinks, and sweetened waters. According to the Centers for Disease Control and Prevention (CDC), 49 percent of adults and 63 percent of young people have at least one sugar-sweetened drink a day. These drinks represent a major health problem in the United States and around the world. **Obesity**, type 2 **diabetes**, and heart disease are **chronic** illnesses that are becoming more common because of the increased amount of sugar in people's diets. Consuming too much sugar can also cause short-term health problems.

Soft drinks can have serious effects when they are consumed too often, and many health organizations are working to help people make more nutritious choices. One way to do this is by making it easier to buy healthier drinks and harder to buy sugar-sweetened drinks. Policies, standards, and laws can influence people's beverage choices—so can education about the harmful effects of soft drinks. Learning what's in your drink and how it might impact your health can help you make these important decisions every day.

Soda can be enjoyed in small quantities and on special occasions, but for some it is an everyday drink.

Chapter One

What's in Your Drink?

Today, soda is more than just water and carbon dioxide. It's also a mix of chemicals, sweeteners, and other ingredients that have no nutritional value. Although they don't add nutrients to your diet, they do add **calories**, and almost all of these calories come from sugar. Beverages make up about 20 percent of the calories children and teens take in every day, so it's important to know what's in the beverages you drink regularly and on special occasions.

Soda Ingredients

Sodas contain dozens of ingredients, which could include sugars and other sweeteners, caramel color, phosphoric acid, caffeine, and natural flavors. The high calorie count and sweetener content have earned soft drinks the name "liquid candy." Most of the individual ingredients in the average can of soda are not nutritious.

The ingredients in packaged foods and beverages are listed on product labels in order of quantity. So the first ingredient on the list is always the most plentiful ingredient. With carbonated beverages, water is the first ingredient. Water is healthy, and humans need it to stay hydrated and survive. However, a form of sweetener is typically listed right after water, usually high-fructose corn syrup, table sugar, or fruit juice sugar concentrates.

High-fructose corn syrup (HFCS) is an artificial sweetener made from corn. Eating this sweetener can increase your hunger and harm your liver over time. High-fructose corn syrup's widespread use in the United States began in the 1970s, and obesity trends have been rising steadily since then.

In addition to HFCS, the added sugars in soda and other soft drinks are listed under many names. Some of these sugars have the term sugar or syrup in them and are easy to recognize. You

When people add spoonfulls of sugar to their coffee or tea, they can control how much of a sweetener they're putting into their drink. The same is not true with soda.

Teaspoons of Sugar

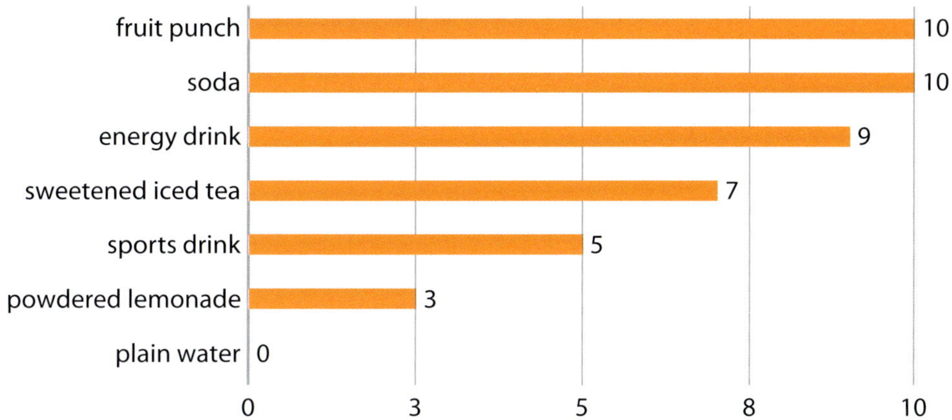

This graph shows the amount of sugar by teaspoon in the average 12-ounce (355 milliliter) container of various beverages. Some soda varieties contain even more than 10 teaspoons of sugar.

can also check the added sugars section of a nutrition facts label to find out how much sugar is in your drink. The American Heart Association recommends that children and teens should have less than 25 grams (0.88 ounces), or six teaspoons, of sugar a day. (Grams are often the standard measurement used when studying foods and drinks.) However, the average can of soda has 39 grams (1.4 oz), or 10 teaspoons, of sugar by itself!

Caramel Color

Another ingredient usually found in soda is caramel color. Like HFCS, caramel color is frequently made from corn. To give carbonated cola beverages their distinct brown color, sugars and carbohydrates such as HFCS are carefully burned with the addition of salts and acids. This process results in a mixture that smells like burned sugar. The caramelized substance gives color to many carbonated beverages, like cola, root beer, and cream soda.

Phosphoric Acid

Phosphoric acid is often the next item in a carbonated beverage's ingredient list. Phosphorus, the source mineral for phosphoric acid, is mined from rocks in the earth. After going through various chemical processes, the phosphorus is converted into phosphoric acid. This acid is quite common and powerful. It is used to make baking soda and can even be used for rust removal.

Phosphoric acid can have a negative effect on the human body. Tooth decay and a higher instance of broken bones have been observed among carbonated soda drinkers. While phosphorus itself is an important part of our bones, a poor phosphorus to calcium ratio may lead to bone loss. Some research shows that kids who drink carbonated beverages often get too much phosphoric acid and too little calcium.

Natural Flavors

So-called natural flavors are another ingredient in many carbonated beverages. The term is vague. The U.S. Food and Drug Administration (FDA) requires that an ingredient identified as a natural flavor must originally come from a natural source, such as a plant or animal material. Consumers generally prefer the term "natural," but most of these flavors are actually produced chemically in the same factories in which artificial flavors are made. The idea that natural flavors are healthier is generally not true.

Drink Water!

When you're thirsty, water is the best drink for your body. Your body loses water as you breathe, digest food, sweat, and remove waste through bowel movements and urination. You have to drink water to keep these processes going.

Water keeps you hydrated and aids all your systems in running smoothly. It helps you regulate your body temperature and keeps you from overheating. That's part of why hydration is so important in hot weather or when you're exercising. Water also makes your joints and tissues smooth and slippery for movement. It keeps your skin healthy, and it prevents painful problems like kidney stones.

Reusable water bottles come in different sizes, shapes, and materials. They can help you avoid single-use plastic water bottles that contribute to plastic waste and pollution.

The amount of water your body needs varies based on body size and activity, but there are general guidelines you can follow. Young people ages 9 to 13 should drink about 7 to 8 cups of water, or 56 to 64 ounces (1.6 to 1.9 liters), a day. People ages 14 to 18 need about 8 to 11 cups, or 64 to 88 ounces (1.9 to 2.6 L), a day. One way to make sure you're drinking enough is with a reusable water bottle. For example, if you had a 16-ounce (0.5 L) bottle, you would fill it four times a day to reach a daily goal of 64 ounces (1.9 L).

Caffeine

Caffeine is an ingredient in many carbonated drinks, especially colas and energy drinks. Caffeine is found naturally in many plants such as the cacao tree, from which chocolate is derived. Caffeine also occurs naturally in coffee beans and tea leaves.

Pure caffeine is a drug that increases the body's heart rate, blood pressure, and other functions of the **metabolism**. When people drink beverages with caffeine in them, they have more energy. While this may give them an initial boost, there is an energy crash soon after.

The American Academy of Pediatrics says that children under 12 should not have any caffeine, and teenagers should have less than 0.004 ounce (100 milligrams) of caffeine a day. It warns against the use of energy drinks for both children and teenagers.

Extra Ingredients

Soft drinks contain other unhealthy ingredients, including artificial colors and artificial sweeteners. Many carbonated beverages contain chemical coloring ingredients like Red 48, Blue 1, Yellow 6, Red 40, and more. The chemical benzene, a flammable oil, is used as the raw material for these food dyes.

Some carbonated beverages are artificially flavored and colored to remind consumers of the natural fruit they imitate. While grape or orange soda may contain no actual orange or grape juice, it is flavored, colored, and labeled to create the idea of a natural fruit source.

The ingredients in soft drinks may not be harmful in small amounts. However, health problems from drinking too many soft drinks can add up over time. There are

Orange soda is made with artificial food dyes, and it's one of the most sugary soft drinks. It has about 13 teaspoons of sugar and 195 calories per serving.

long-term and short-term effects of drinking soft drinks for people of all ages.

Food for Thought

1. What are the benefits of knowing what's in your drink?
2. Are there any ingredients that you regularly check for on a nutrition facts label, such as high-fructose corn syrup?
3. What can you learn from a nutrition facts label besides the ingredients?

A 20-ounce (591.5 mL) bottle contains multiple servings of soda. If you drink the whole bottle, you're getting more than double the sugar of a single serving.

Chapter Two

Immediate Effects

Beverages are the largest source of sugar in American diets. Although sugar tastes delicious, a lot of it is difficult for the body to process. Drinking sweetened soft drinks on a regular basis can change the way your body functions over time and lead to chronic diseases.

Not all the effects of sugary beverages happen in the distant future, though. These drinks can also have an immediate or short-term impact on your body. Drinking too many soft drinks can cause problems with your energy, mood, bones, and teeth.

To avoid health problems, the American Heart Association recommends that 2- to 18-year-olds have no more than one 8-ounce (236 mL) sugar-sweetened drink a week, yet most young people have more than that amount each day. In addition, most soft drinks are sold in containers that hold more than 8 ounces (236 mL). Cans usually have 12 ounces (355 mL), plastic bottles hold 20 ounces (591.5 mL), and family size bottles contain 68 ounces (2 liters) or more.

Energy and Rest

The quickest effects from sugary drinks are energy rushes followed by crashes. The sugar, caffeine, and calories in most soft drinks may give you a jolt of energy, but not for long. Soon you will feel sluggish as your metabolism works harder to process the extra sugar. In addition to affecting your energy level, sugary drinks can also have a negative effect on your mood.

Poor sleep patterns are an immediate side effect of drinking too much soda. The ups and downs created by caffeine and sugar keep the body from resting properly. This leads to sleepiness and a lack of concentration. Sleep is necessary for the body to heal and make new, healthy cells. Chronic lack of sleep blocks this natural healing process.

Caffeine Addiction

Caffeine is a stimulating drug that increases heart rate and blood pressure. People who consume caffeinated soft drinks on a regular basis become used to caffeine and have uncomfortable symptoms when they don't have it. Irritability, headaches, nausea, poor concentration, and sleepiness are all symptoms of caffeine **withdrawal**.

Caffeine is addictive, and its power lessens as your body gets used to it. That can make you drink more of a caffeinated beverage over time. If you get caffeine from soda, you're also getting sweeteners, artificial colors and flavors, and other chemicals. Sugar and caffeine can make you swing between high energy and sluggishness, affecting your mood and ability to concentrate.

Caffeine isn't only in soda. It's also found in energy drinks and some sports drinks. Although sports drinks are designed to support healthy athletes, they can still have high amounts of calories, sugar, and caffeine.

Sports drinks can help you stay hydrated when you're active on a sports team or moving a lot, but they should be saved for days when you're active for more than 1 hour, and they should be caffeine free.

Should You Drink Diet Soda?

Diet soda sounds like it would be better for you than regular soda. However, diet soda is not a healthy beverage either. Diet sodas use artificial sweeteners made from chemicals in order to keep the calorie count low. There might be short-term weight loss from switching from regular soda to diet, but there are no health benefits in the long run. All the long-term risks of regularly drinking diet soda aren't fully clear yet. However, it is clear that drinking diet soda does not lower the risk of developing type 2 diabetes. Drinking diet soda also increases the risk of heart disease and **stroke** in women.

Some people use diet soda to help them drink less regular soda. Once they've switched to diet, the next step is to switch again to something healthier. If you like a fizzy drink, plain carbonated water is the best choice. Many flavored sparkling waters are available, too, and some are free from added sweeteners. To protect the enamel on your teeth from citric acids, flavored sparkling water is best enjoyed with a meal or with a glass of still water.

When you need a boost, it's best to get energy from a healthy snack or a glass of water or low-fat milk. Water hydrates and gets rid of toxins, resulting in higher sustained energy levels. Milk helps build stronger bones, provides steady levels of energy, and creates a feeling of fullness. While plant-based milks are often better for the planet, they can have more sugar and fewer nutrients kids need than cow's milk.

Teeth and Bones

Tooth decay is a result of too much sugar in the diet. Sugars from beverages can linger in the mouth and mix with acids in saliva, or spit. Over time, this sugary acid wears down the enamel, or

Soda is acidic and can erode the enamel on teeth more quickly than other drinks, especially if a person regularly drinks sugary beverages.

Although there are calories in milk, they come from helpful nutrients instead of just from sugar.

protective layer, on teeth and causes cavities. Untreated cavities can lead to tooth rot, dangerous infections, and bad breath. Tooth decay can happen rather quickly, rotting teeth in a matter of weeks or months.

Tooth decay and gum disease can also result from not having enough milk and calcium. Teeth require milk and calcium to stay healthy and strong, and so do bones. Doctors have seen increased broken bones in children under the age of 18. This could be because children and teens are drinking less milk than they should. An unhealthy weight that is higher may also be a factor as it places stress on bones and joints. Milk is an important source of vitamins D, A, and B12, which are essential nutrients that soft drinks do not provide. Many soft drink companies try to add nutrients to their products, but these added nutrients are not metabolized as well as natural sources. Fresh fruit or vegetable juices and milk are the best beverages for getting the vitamins and minerals your body needs.

Food for Thought

1. How can you raise your energy level without having caffeinated or sugary drinks?
2. Do you think it's helpful to switch to diet soda in order to cut back on regular soda?
3. Do you notice changes in your mood when you drink sugary beverages?

Sugar-sweetened drinks can play a part in the development of many health concerns.

Chapter Three

Long-Term Problems

Having too many sugary soft drinks can lead to unhealthy weight gain. Your body turns extra sugar into fat and stores it. In addition, sweet drinks are often high in calories, but they don't make you feel full the same way food does. Taking in too much sugar and too many calories can lead to gaining an unhealthy amount of weight.

A study shows that children who drink 12 ounces (355 mL) of soda a day are 60 percent more likely than others to live with obesity. Childhood obesity has increased in the United States, practically doubling between 1988–1994 and 2017–2018. It shot up again in 2020. In that year, 22.4 percent of children were living with obesity. As rates of this condition increase, so do the rates of related diseases like type 2 diabetes, liver disease, heart disease, and **cancer**.

Diabetes

Diabetes is a disease that causes high blood sugar levels because the body either doesn't produce enough **insulin** or the body's cells don't respond to the insulin that is produced.

Childhood Obesity in the United States

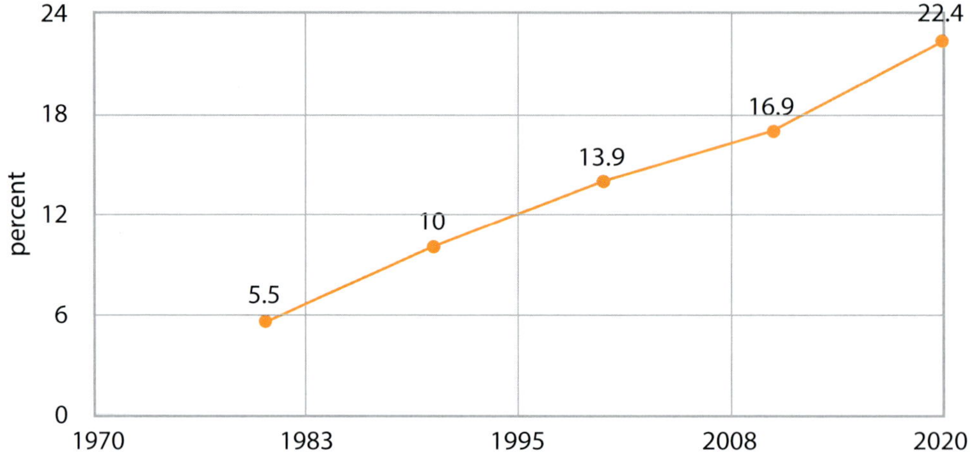

The rate of childhood obesity has been climbing in the United States, reaching a high of 22.4 percent in 2020.

Insulin regulates the amount of **glucose** in the blood. Glucose is created by the body during digestion. It is absorbed by the body's cells and provides them with energy.

Type 1 diabetes is caused by an immune system problem that results in the body's failure to produce insulin. Type 2 diabetes used to affect mostly adults because its onset is more closely related to food and lifestyle habits, but it has been increasing in children ages 10 to 19. With type 2 diabetes, the body's cells become insulin resistant and do not absorb glucose effectively, leading to high blood sugar levels. The Cleveland Clinic has shown that drinking one or more sodas a day increases the risk of developing type 2 diabetes by 67 percent.

Drinking sugary drinks too frequently will keep the pancreas working too hard for too long because it is constantly trying to regulate glucose levels. Over time, the body will become resistant to all the excess insulin. Once the body's cells become insulin resistant, blood sugar levels shoot up. Eventually, the pancreas wears out and cannot keep up. The body can no longer produce or absorb enough insulin to regulate glucose once type 2 diabetes develops. The consequences can be

According to the CDC, in 2019, about 283,000 children and teenagers had diabetes. People with diabetes often check their blood sugar using a glucometer.

severe—vision loss, nerve damage, cardiovascular disease, kidney disease and failure, coma, and even death.

Many people with type 2 diabetes live with obesity. There are other early warning signs that diabetes might be developing. These include extreme thirst, frequent urination, sudden changes to vision, an increased appetite, sudden weight loss, and unusual drowsiness. Anyone experiencing these symptoms should see a doctor immediately.

Cirrhosis

Cirrhosis is a chronic form of liver disease. Cirrhosis can result in fluid **retention**, bone pain, shrinking or swelling of the liver, yellowing of the skin and eyes, coma, and death. Researchers have found that some cases of cirrhosis are caused by a preservative in sodas called sodium benzoate.

Sodium benzoate is made from benzoic acid. This acid can be found in small amounts in nature. However, sodium benzoate is chemically produced in large quantities and put into sodas to prevent mold from forming in the can or bottle.

In addition to cirrhosis, scientists fear that sodium benzoate may cause cancer when consumed over a long period of time. When the preservative is mixed with vitamin C, which is found in many soft drinks, a cancer-causing substance called benzene is produced. Sodium benzoate is also linked to the development of **Parkinson's disease**.

Heart Disease

Diabetes and obesity put people at an increased risk for developing heart disease. Higher levels of cholesterol are produced when a person lives with obesity, and this increased fat in the blood can eventually clog arteries. Strokes are related to both high blood pressure and diabetes. Diabetes, high blood

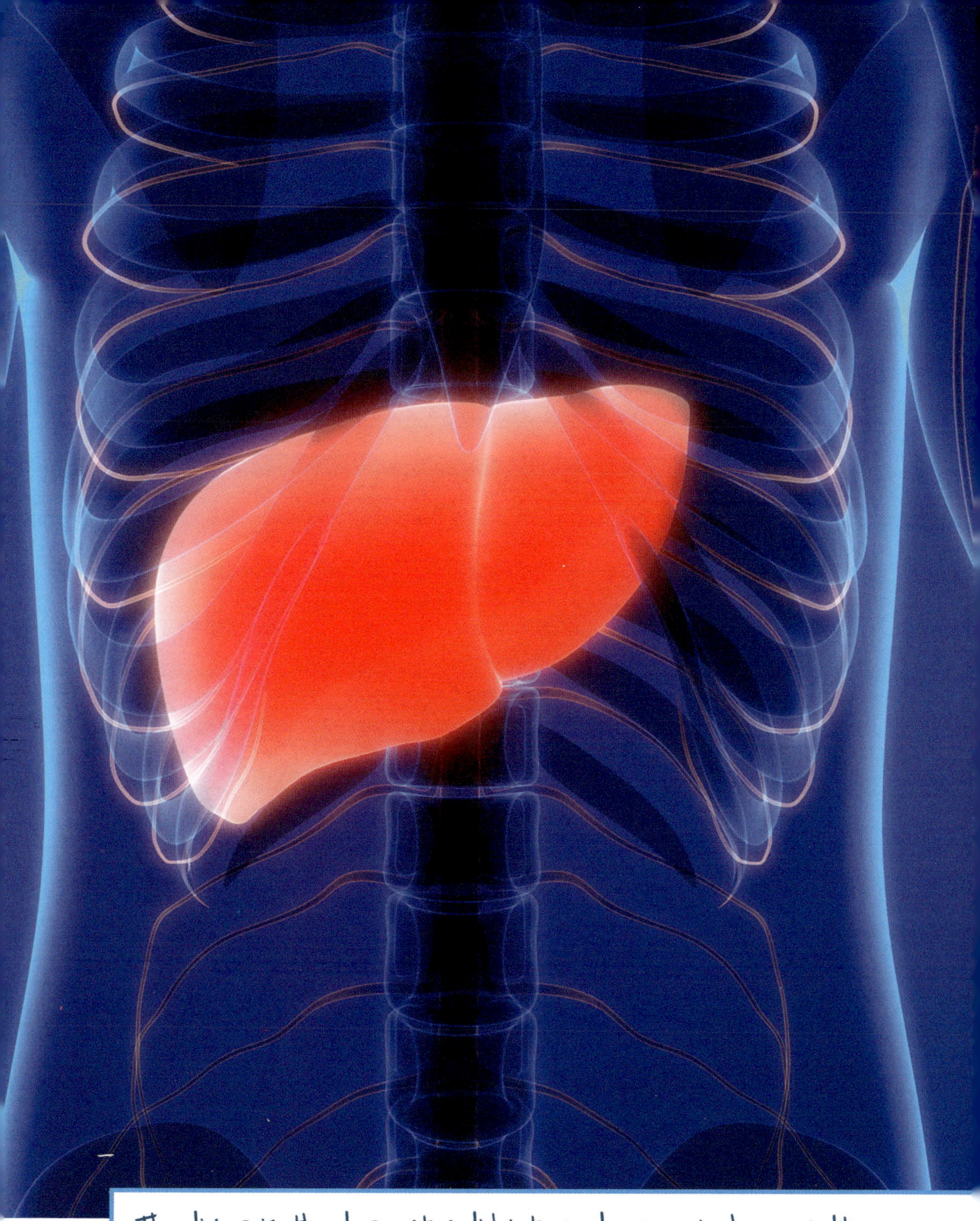

The liver is the largest solid internal organ in humans. It helps the body filter blood, create chemicals, make bile for breaking down food, and use nutrients from food.

pressure, and being overweight can all lead to stroke and heart disease over time.

People who drink soda every day have a 36 percent higher risk of developing metabolic syndrome, a common cause of heart disease. Metabolic syndrome often occurs in people who have an unhealthy weight (especially if there's extra weight at the waistline), have high blood pressure, have high blood sugar, and have high cholesterol.

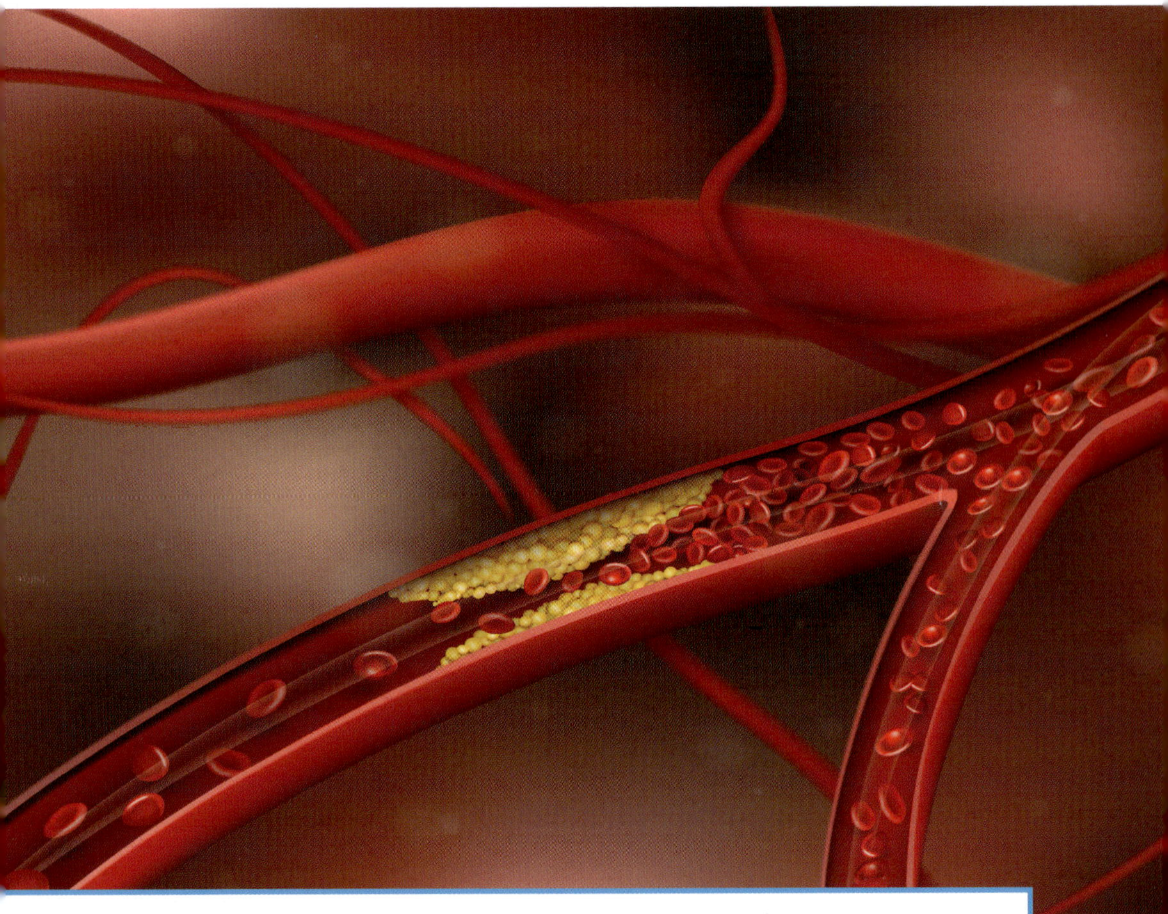

Having high cholesterol causes plaque to build up on the walls of arteries. In this model, plaque is the yellow substance that restricts blood flow through an artery.

Cancer

Cancer is closely related to chronic diseases like obesity, type 2 diabetes, and heart disease that can result from frequently drinking sugar-sweetened beverages. Certain types of cancer have been linked with soft drinks. For example, a study at the University of Minnesota found that people who drink as few as two sodas per week are twice as likely to develop pancreatic cancer as people who do not. The pancreas is the place where insulin is produced in an effort to balance and regulate the body's blood sugar level. If the pancreas is working extra hard, it may cause an abnormal growth of pancreatic cells, which results in cancer.

Regularly drinking large amounts of carbonated beverages can cause an upset stomach and bloating. Some people get heartburn or **acid reflux** from soda consumption. Acid reflux happens when stomach acid is sent back up from the stomach to the **esophagus**. People with acid reflux become more likely to develop esophageal cancer because of damage to the esophagus.

The stomach is lined with a protective coating that keeps digestive acids from digesting the stomach itself. When too much acid from soft drinks is deposited in the stomach on a consistent basis, it can wear away the protective lining, causing small burns and holes. These are called ulcers. Some ulcers can become cancerous.

A Public Health Problem

Sugar-sweetened beverages are a major public health problem since the effects of drinking them are so serious. Experts have many ideas for how to encourage healthier drink choices. One way is by establishing soda taxes. The World Health

Plastic from bottled beverages is polluting Earth's oceans. Drinking less soda has an environmental benefit since you'll be creating less plastic waste. Instead of buying bottled water, try filling a reusable bottle with tap or filtered water.

Organization (WHO) and other public health institutions recommend that governments tax sugar-sweetened beverages like soda.

Supporters of the tax argue that soda is not food: It has no nutritional value and is known to be harmful to health. They think it should be taxed just as tobacco and alcohol are. They hope that raising the cost of soda through taxes will discourage people from buying it.

A soda tax was introduced in Mexico in 2014. Within the first year, the sale of sugar-sweetened beverages decreased by 12 percent. Although the U.S. federal government hasn't adopted these recommendations, some state and city governments have. Cities with soda taxes include Philadelphia, Pennsylvania; the District of Columbia; Boulder, Colorado; Seattle, Washington; and Albany, Berkeley, Oakland, and San Francisco in California. Berkeley saw a drop in soda sales and an increase in bottled water sales as a result of their tax, which was passed in 2015.

Experts from the Harvard T. H. Chan School of Public Health also recommend changing labeling requirements. They believe labels on sweetened drinks should warn of the connection with obesity and diabetes, similar to the warnings found on tobacco and alcohol products. Any drink with more than 50 calories in an 8 ounce (236 mL) serving would receive this label.

Regulating the marketing of sugary drinks to kids is another way the government can address America's unhealthy soda habit. This would mean placing limits on advertisements for sugar-sweetened beverages that target kids on the internet, watching TV, and engaging with other media.

Some cities have already acted to curb sugary drinks. The city of San Francisco in California has taken on the advertisement

Soda Tax in Action

A tax on sweetened beverages has been in place in Philadelphia, Pennsylvania, since 2017. Philadelphia was the first big U.S. city to adopt a soda tax. The city pledged to use the extra 1.5 cents per ounce from the tax to support early education. In 2021, $70 million was generated by the tax.

Data from the first year of the tax in Philadelphia recorded a drop in sweetened beverage sales by 38.9 percent, but the overall impact was mixed. A study by Drexel University recorded few changes in the beverage habits of Philadelphians. Instead, many people continued drinking sweetened beverages but purchased them outside the city to avoid the tax. However, another study showed that high schoolers in Philadelphia drank less soda than teens in other cities.

Opposition to Philadelphia's tax has been strong. Increased costs can affect people with low incomes who choose to keep buying sweetened drinks despite the higher prices. Although the city benefits financially, the tax has not had a notable impact on obesity. To have the greatest effect on health, soda taxes must be part of a larger public health approach to educate people about harmful habits and encourage change.

of sugary drinks by placing health warnings on billboards and large outdoor ads for sodas. Boston, Massachusetts, has developed a number of restrictions on sugary drinks, including a ban on selling them in school vending machines and city-owned buildings. New York City tried to pass a ban on sodas larger than 16 ounces (473 mL) in 2013, but it was struck down in court.

Philadelphia's sales tax applies to any beverage with added sugar or sweetener, including energy drinks, sports drinks, fruit punch, diet soda, and regular soda.

Food for Thought

1. Should the government be able to ban or tax beverages with harmful health effects? Why or why not?
2. Do you think there should be warning labels on sugar-sweetened drinks?
3. Does knowing the long-term health effects of drinking sugary beverages change the way you think about them?

If you're tired of plain water, flavor it yourself by chilling fresh fruits, vegetables, or herbs in a jar or pitcher of water.

Chapter Four

Healthy Choices Every Day

The chronic diseases that result from consuming too many sweetened soft drinks affect individuals, but they also have an impact on whole communities and countries. Diet-related health problems are emotionally and physically painful. They are also expensive. In the United States, $173 billion in health-care costs are paid every year as a result of obesity. There are many diseases that can't be prevented, but you have a higher chance of avoiding some diseases by taking charge of what you eat and drink. Fortunately, there are many alternatives to soft drinks that are not only good for you, but also good tasting.

Rethinking Soft Drinks

Today, soft drinks are so readily available that they are often thought of as an essential part of meals and snacks. If people return to the idea that sweetened beverages should only be consumed on special occasions—like parties and cookouts—they could greatly decrease their daily intake of sugar and chemically processed preservatives.

Try adding fruit or a splash of 100-percent fruit juice to unflavored sparkling water. The drink will be sweet without adding too much sugar to your day.

Ways to Change

Soda is a big threat to health in the United States and it affects every community. In addition to government actions to regulate soda, there are things individuals, companies, schools, and workplaces can do to work together and help people make healthy drink choices. To start, individuals can learn about the health issues that can come from drinking soft drinks and try to change their habits every day.

The companies that make sweetened drinks can also design their product to have less sugar or sweetener and manufacture smaller bottles and cans. The Harvard School of Public Health suggests that drinks should have no artificial sweeteners, contain no more than 1 gram of sugar per 1 ounce (29 mL), and come in 8 ounce (236 mL) bottles.

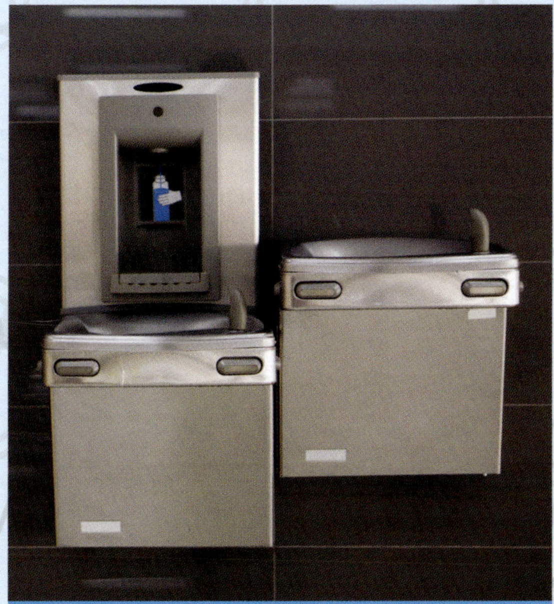

Schools can encourage healthy hydration by maintaining working drinking fountains where students can get a drink or fill a reusable bottle.

Schools and workplaces can encourage healthy habits by making sure filtered water or drinking fountains are available and in working condition. They can also decide not to sell sugary drinks and sodas in vending machines and cafeterias. The Smart Snacks in School standards require all food and drinks sold in U.S. public schools to meet certain conditions. Based on these standards, sugary, carbonated drinks should not be sold in elementary and middle schools.

Everyday Drinks

Milk is one of the most nutritious beverages. It provides calcium that the body needs to grow and stay strong, it is loaded with vitamins A and D, and it helps create a full feeling that prevents overeating. Children ages 9 to 13 should have about three servings of dairy a day, and a glass of milk can make up a serving. Most nutritionists recommend drinking low-fat or nonfat milk. People with milk allergies can look for dairy free options that have vitamins but don't have added sugars.

Water is always an excellent beverage choice. The body needs water to maintain all of its functions, and it helps to metabolize food. Best of all, it has no added sugar, salt, or artificial chemical coloring, flavoring, or sweeteners. Keep a bottle of water in the fridge for a refreshing treat in place of soda. For an extra flavor kick, a wedge of lemon, lime, or orange can be squeezed in.

Flavors and Fizz

Some people enjoy the carbonation of sodas. They could try carbonated or sparkling water instead. Check the nutrition and ingredient labels on these beverages, however. Make sure they don't include extra calories, added sugar or artificial sweeteners, or too much sodium.

Fruit juices can also serve as a healthier alternative to sodas. However, many juices have artificial colors and flavors. Some do not even include real fruit. Even fruit juices made with natural fruits can have added sugars and high-fructose corn syrup. To avoid these sugary options, look for drinks that are 100 percent juice.

Juice contains a lot of natural sugar, but it also has nutritional value. Be sure to choose 100 percent juice without added sweeteners.

Fresh-squeezed juices with nothing added are always the best choice. Consider making juices at home with a blender or juicer. This way, you can control what goes into the drink and come up with delicious fruit and vegetable combinations. Even fresh-squeezed juices are high in sugar and calories, so it's best to only have a small amount of juice.

You can make healthy beverages part of your family meals and get in the habit of saving soft drinks only for rare treats.

Check the Label

When choosing beverages, be sure to check ingredient lists and nutrition labels. Typically, the healthier choices have fewer ingredients, less sugar, less sodium, no high-fructose corn syrup, no caffeine, fewer calories, and no artificial colors, flavors, or sweeteners. A long ingredient list of chemicals that are difficult to pronounce is a sure sign that a beverage is not a healthy choice. Making better beverage choices will help lead to a lifetime of good health.

Food for Thought

1. What fruits, vegetables, or herbs might you try adding to water?
2. How much control do you have over what you drink?
3. What small steps can you take toward making healthy beverage choices?

Glossary

acid reflux: A condition that causes the regurgitation of stomach acids into the esophagus.

calorie: A unit of heat energy that animals get from food.

cancer: A disease caused by abnormal cell division in a certain part of the body.

carbonated: Having a bubbly, fizzy quality that results from dissolving carbon dioxide in a liquid.

chronic: Always present, as in a long-lasting disease or condition.

diabetes: A disease in which the pancreas does not produce any or enough insulin or the body's cells cannot process the insulin that is produced.

esophagus: The muscular tube through which food passes on its way to the stomach.

glucose: A simple sugar that is needed for energy in living things.

insulin: A hormone made by the pancreas that regulates glucose levels in the blood.

metabolism: The body's processes, including ways it uses food and water to make energy.

obesity: A condition in which an unhealthy amount of fat is accumulated and stored in the body.

Parkinson's disease: A progressive disease of the nervous system that causes muscular tension and tremors.

retention: The process of keeping extra liquid, heat, or another substance inside the body.

stroke: A blockage that can cause damage by preventing blood and oxygen from flowing to an area of the brain.

withdrawal: Physical or mental symptoms as a result of limiting intake of an addictive drug.

Find Out More

Books

Benson, Jodyanne. *Diabetes*. New York, NY: Rosen Publishing, 2021.

Chandler, Matt. *Understanding Obesity*. Ann Arbor, MI: Cherry Lake Publishing, 2020.

Gallagher, Jamie. *The Chemistry of Cola*. Clayton, Australia: CSIRO Publishing, 2020.

Websites

Britannica Kids
kids.britannica.com/students/article/soft-drink/277126
This encyclopedia article explains the history of soft drinks and how they're made.

Centers for Disease Control and Prevention
www.cdc.gov/healthyweight/healthy_eating/drinks.html
This page from the CDC defines sugar-sweetened beverages, offers tips for finding the sugar content of drinks, and suggests healthier alternatives to soft drinks.

Publisher's note to educators and parents: Our editors have carefully reviewed these websites to ensure that they are suitable for students. Many websites change frequently, however, and we cannot guarantee that a site's future contents will continue to meet our high standards of quality and educational value. Be advised that students should be closely supervised whenever they access the internet.

Organizations

Action for Healthy Kids
600 West Van Buren Street, Suite 720
Chicago, IL 60607
www.actionforhealthykids.org
Action for Healthy Kids works throughout the country to educate people on the importance of healthy eating and exercise and to reduce childhood obesity.

American Diabetes Association
2451 Crystal Drive, Suite 900
Arlington, VA, 22202
www.diabetes.org/
The American Diabetes Association aims to prevent diabetes through education, supports people diagnosed with diabetes, and works to find a cure.

Whole Kids Foundation
550 Bowie Street
Austin, TX, 78703
www.wholekidsfoundation.org/
This is a nonprofit organization that educates kids about healthy eating and supports good nutrition at school and at home.

Index

About the Author

Rachael Morlock is a freelance writer and copyeditor. She is the author of several nonfiction and picture books for children and enjoys researching new subjects through her work. In the past, she worked as the coordinator of a Community Supported Bakery. Rachael is always interested in supporting local farmers and food producers in Western New York, where she lives with her dog, Rilke.